Cool Careers in
MEDICAL SCIENCES

Sally Ride
Science

CONTENTS

Rhea

Cosette

Carlos

Donald

Gina

Vincent

Miriam

Edith

Ralph

Benjamin

Julie

Deborah

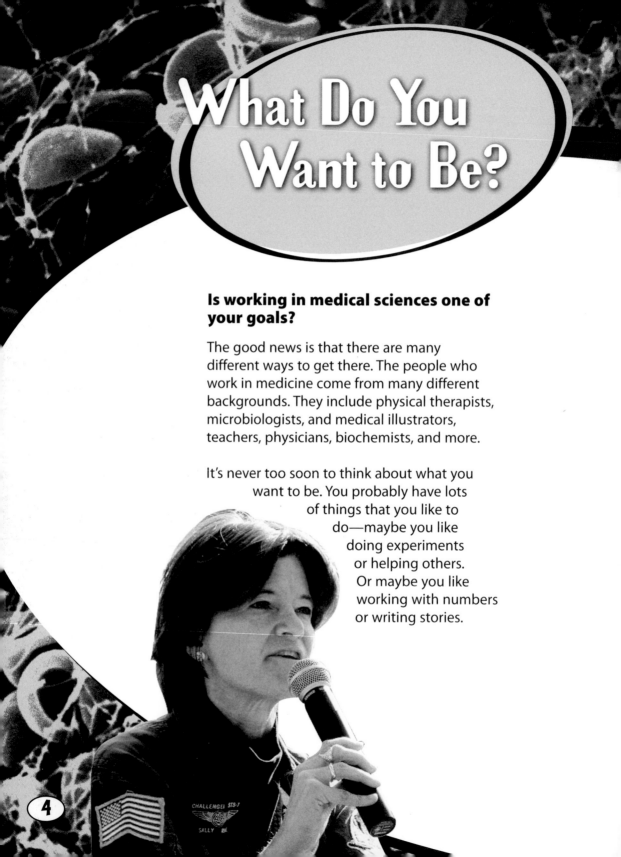

What Do You Want to Be?

Is working in medical sciences one of your goals?

The good news is that there are many different ways to get there. The people who work in medicine come from many different backgrounds. They include physical therapists, microbiologists, and medical illustrators, teachers, physicians, biochemists, and more.

It's never too soon to think about what you want to be. You probably have lots of things that you like to do—maybe you like doing experiments or helping others. Or maybe you like working with numbers or writing stories.

SALLY RIDE
First American Woman in Space

The women and men you're about to meet found their careers by doing what they love. As you read this book and do the activities, think about what you like doing. Then follow your interests, and see where they take you. You just might find your career, too.

Reach for the stars!

Sally K Ride

RHEA SEDDON

LifeWings Partners, LLC

On a Medical Mission

As a girl, Rhea Seddon watched the first American astronauts blast into space. She dreamed of becoming one. But that was in the 1960s, and NASA did not allow women to become astronauts. So she pursued another interest—medicine. Rhea became a surgeon. Her hopes of going into space didn't fade, though. When NASA opened the astronaut program to women, she applied. In 1978, Rhea was one of the first six women chosen to become an astronaut. She went on to fly three missions aboard the Space Shuttle.

Doc in Orbit

As an astronaut, Rhea put her medical smarts to good use. She conducted medical experiments on the ways the human body changes in space. She studied how the body is affected by weightlessness. It can weaken the muscles and bones and the body's immune system. She performed exercise tests and wore special monitoring equipment. Weightlessness can also lower the number of blood cells in a person's body. So, she took samples of her own blood and brought them back to Earth to be analyzed.

Aboard the Space Shuttle, Rhea breathes into an instrument attached to an exercise bike. It measures how well her heart and lungs work in space.

An astronaut is trained to go into space to explore our world and beyond. Astronauts represent many different professions and areas of expertise. Rhea is a physician who studied the human body in space. Other **astronauts**

✴ conduct weightlessness experiments on crystals, insects, or plants.

✴ release satellites into space.

✴ study Earth from space.

✴ pilot the Space Shuttle.

Rhea floats in weightlessness in NASA's research plane.

Around and Around

The Space Shuttle orbits Earth at about 28,000 kilometers (17,500 miles) per hour. It circles our planet once every 90 minutes. That's 16 trips around the globe each day! Imagine you're on an eight-day Shuttle mission.

• How many times would you loop around Earth during the mission?

• How far would you travel each day, in kilometers? In miles?

About You

Rhea's prescription—"Get as much education and as good an education as possible in a technical field."

Team up with a classmate and discuss what area of science or engineering interests you and why. Research and find a career in this area that interests each of you.

Helping the Next Generation

Rhea is helping NASA study ways to keep astronauts healthy during a long space mission, such as a journey to Mars. Now it's your turn to help. Make a list of some things Rhea needs to think about and why they're important.

Check out your answers on page 36.

COSETTE SERABJIT-SINGH

GlaxoSmithKline

"Science is a great profession that actively encourages you to keep learning. It also gives you a pattern of thinking to help you do that."

Good Science, Better Medicine

The next time you feel better after taking some medicine, thank someone like Cosette Serabjit-Singh. She's a biochemist who works for a pharmaceutical company—a company that makes medicines. She works toward making drugs safe and effective by studying what happens to them once they enter a person's body.

Hands-on Happiness

Cosette likes to help discover medicines for people in need. She oversees a team that is studying new drugs to treat patients with breast cancer, diabetes, and other diseases. She also likes working in her laboratory. There she works with other chemists. Together they come up with new research questions and conduct experiments. Working with her hands to solve problems is what drew Cosette to science as a young girl growing up in Jamaica. "Science class was the one place where you got to *do* something!" she says.

"In science, coming up with a really good question can be more important than coming up with the right answer."

A biochemist studies the chemistry of living organisms. Cosette works for a pharmaceutical company. Other **biochemists**

✳ work in hospital laboratories, where they study tissues that can help them understand and treat disease.

✳ work in the food industry, where they check the safety of food and beverages.

✳ work in a university, where they study chemical reactions that build bones, fight infections, or store memories.

Stick to It

Have you ever been stumped by a tough problem? Cosette has some advice for you. "Find your inner voice that says, 'I bet this is not that hard if I just stop to think about it.'"

Is It 4 U?

What parts of Cosette's work do you like, and why?

- Asking questions
- Investigating problems
- Working on a team
- Conducting hands-on experiments
- Giving talks at scientific conferences
- Writing research papers
- Helping people

Team up and discuss what you would like about being a biochemist.

Lab Work

Biochemists use special tools when they work in the laboratory. Research how they use each of these tools and create a chart with three columns. In the first column write down the name of the tool. In the second column, describe some of the things it's used for. In the third column, draw a picture of the tool.

- Flask
- Pipette
- Test tube
- Microscope slide
- Safety goggles
- Spectrometer

Math Problems Solve Problems

Real-world results make it easy for Carlos to talk to young people about the importance of math and science. "I tell them that with math and science you can actually address the problems that you're interested in. And, you can make a difference."

CARLOS CASTILLO-CHAVEZ

Arizona State University

Math As a Cure?

When a terrible disease breaks out, Carlos Castillo-Chavez starts crunching numbers. He is a mathematical epidemiologist. He uses math to study the way diseases such as malaria, influenza, and chicken pox get started. He also uses math to calculate the time they will take to spread, and the best way to stop them.

Calculating a Disease

Following an outbreak in Canada of a deadly respiratory illness called SARS, Carlos started his number crunching. He used a mathematical model to figure out just how quickly the disease was being passed from person to person, and how to put an end to it. His model confirmed that Canadian health officials were effective in stopping the outbreak. They kept people infected with SARS away from others. They also shut down two hospitals where the virus was being transmitted. Carlos and his students determined that these actions may have prevented as many as 200,000 people from getting sick. They also correctly predicted that about 400 people in Canada would become ill because of SARS.

Carlos enjoys encouraging others interested in math, such as this group of students.

An epidemiologist studies outbreaks of diseases that affect many people at the same time or in the same area. Carlos uses math to analyze outbreaks. Other **epidemiologists**

* travel to sites of infectious illnesses to determine how an epidemic started.
* work with health officials to control outbreaks.
* encourage healthful lifestyles and behaviors to prevent diseases.
* study environmental factors that might expose people to illness.

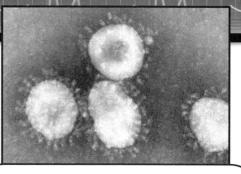

A strain of the coronavirus was responsible for the spread of SARS. When viewed under the microscope, this virus has a halo, or corona.

Ask Around

Growing up in Mexico City, Carlos dreamed of becoming a hotel manager. He thought it sounded glamorous. He later worked at a cheese factory, where the tiring physical work motivated him to go back to school to finish his education. Ask a grown-up whom you respect what they wanted to do at your age. How did they get to the point where they are today?

Funny Bone

Q. Where did the math student eat her lunch?

A. At the multiplication table, of course!

You, the Epidemiologist

A bad case of food poisoning has broken out among people who went to the county fair. Why? Study the facts below, then write your hypothesis, or educated guess.

1. Those who were sick all ate barbecued chicken from the Cheap Cheep Chicken Stand.

2. Not everyone who ate barbecued chicken at the fair became ill.

3. Everyone who was sick ate lunch after 1 P.M. Those who didn't get sick ate their chicken before 1 P.M.

4. The chicken stand's electric grill wasn't working well on the afternoon that everyone became sick.

Check out your answers on page 36.

In Don's job, no two days are alike. One day he could be illustrating a certain disease, and the next he could be drawing a brain in 3-D.

Explaining Science

After taking lots of medical and art classes in college, Don went on to become a medical illustrator for the nation's top research center in medicine. Through drawings and animated videos, he helps to explain the research that scientists are doing.

DONALD BLISS

National Institutes of Health

Serious Art

Don Bliss was planning to become a doctor until he noticed a poster outside a biology professor's door. It listed all the careers you can pursue with a biology degree. One was medical illustration—creating art that explains scientific and medical concepts. Don was excited. He'd loved drawing since he was a young boy. He'd never thought it was something he could do for a living. As a medical illustrator, he'd get to combine art and biology, his two favorite subjects.

Worth a Thousand Words

Don starts his illustrations on paper, then he scans the drawings into a computer to finish them using a software program. Scientists are often thrilled to see their hard work simplified. "One of the most satisfying parts of my work is helping scientists to communicate something that's really hard to explain with just words," he says.

Don's drawings help explain medical concepts, such as how the brain communicates with other parts of the body.

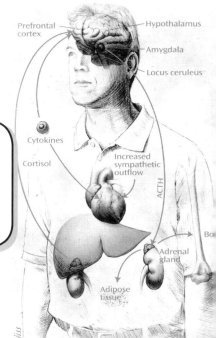

Prefrontal cortex
Hypothalamus
Amygdala
Locus ceruleus
Cytokines
Cortisol
Increased sympathetic outflow
ACTH
Adrenal gland
Adipose tissue

A medical illustrator is

a professional artist who uses art to explain medicine. Don illustrates everything from what nerve cells look like when a person is in pain to how DNA unwinds to make a copy of itself. **Medical illustrators** also

* illustrate medical textbooks or wall charts for doctors' offices.

* work at medical schools or large hospitals.

* create animated films using computers to show how the human body works.

Heart Art

What do you know about the human heart? Do a little research. Then grab some colored pencils or markers. Draw a heart and show how blood flows through it. Even though blood is all the same color—red—illustrators usually use blue to show blood flowing into the heart. They use red to show blood flowing out of the heart. Write a caption that explains how blood flowing into the heart is different from blood flowing out of it.

About You

Don liked painting and drawing as a boy—and still does. What are your favorite things to do outside of school?

Try Your Hand

Don took many of the same classes as medical students—he even dissected a cadaver! Imagine all the details medical illustrators need to know in order to create accurate drawings of the human body. Here are just a few.

* You have about 100,000 hairs on your head.
* There are about 100 billion nerve cells in your brain.
* You use about 13 muscles to smile and about 33 to frown.
* There are about 206 bones in your body.

While you're looking at the top of one of your hands, draw it with as much detail as possible. Share and compare your drawing with a classmate. What details make each drawing realistic? What details could you add?

GINA KOLATA

The New York Times

The Write Stuff

How do you become a top-notch science and medical reporter? For *New York Times* journalist Gina Kolata, the key is having a background in science and math. Gina earned a college degree in biology, then started a Ph.D. program in molecular biology. But she realized that what she really enjoyed was reading about the latest scientific discoveries and then writing about them. A career as a medical reporter made sense, especially because she had loved writing since she was a girl.

Science Scoops

As a medical journalist, Gina uses her science smarts. She figures out which medical breakthroughs, discoveries, and controversies will make good news stories and which ones won't. In her stories, Gina tries to capture the excitement or frustration of the scientists she interviews. She wants her readers to appreciate their work. She says she's doing her job if readers want to read her stories all the way to the end.

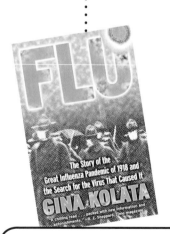

FLU

The Story of the Great Influenza Pandemic of 1918 and the Search for the Virus That Caused It

GINA KOLATA

Gina has also written five books on science, including one on physical fitness and one on a famous flu epidemic.

A medical reporter

writes and reports on science and medical news. Gina works for a daily newspaper. Other **medical writers**

✳ work for hospitals or medical schools.

✳ write medical novels.

✳ write for science magazines.

✳ report medical news on television, radio, or online.

No to B.O.

Suppose there's a new discovery of a way to get rid of body odor forever. Sweet! Imagine you're a reporter for your school newspaper, and this is a hot story. Your assignment is to write about it.

Take a crack at writing the first few sentences of your story—journalists call this the "lead." Use the lead to grab the interest of your readers.

About You

Gina likes reading and writing nonfiction, such as news, history, biographies, and how-to books. Do you prefer to read nonfiction or fiction? Why?

Stack of Papers

There are more than 1,000 daily newspapers published in the U.S. *The New York Times* is one of the biggest. It sells about a million copies a day! Do you think this number will get larger, smaller, or stay the same in the future? Why?

In Your Life

Pick a partner and together find an article in a newspaper or magazine about health. After you each have read the story, take turns answering these questions.

- What was the article about?
- What was interesting about the article?
- What did the story have to do with your health, or the health of someone you know?
- Was there anything the writer could have explained better?

VINCENT PIERIBONE

Yale University School of Medicine

Scuba Science

Would you like a job that includes scuba diving? Welcome to Vincent Pieribone's world. As a neurobiologist, he does cutting-edge research on the way the brain works. This research involves studying the cells of the nervous system. Sure, much of Vincent's work takes place in the lab, but he also works in some of the most beautiful spots in the world. Imagine your lab being Australia's Great Barrier Reef.

Spying on Cells

Vincent searches coral reefs for a type of fluorescent protein that he uses in his research. By inserting copies of it into the cells of living animals, Vincent can make the cells glow. He can then watch them as they divide, grow, and generally go about their business. The technology is still in its early stages of usefulness, but Vincent hopes it will one day help unlock the mystery of how our brain cells work. For example, what steps do our nerve cells take so that a thought, like "I'm thirsty," becomes an action, like taking a sip of water? Unlocking these mysteries will help people who are paralyzed.

Vincent started a research company that he hopes will produce medicine from the sea.

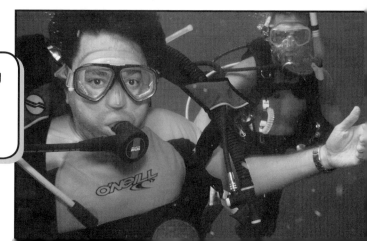

Vincent loves scuba diving in coral reefs and cares a lot about protecting these fragile environments.

A neurobiologist is a scientist who studies the nervous system. Vincent investigates how messages in the brain result in movements throughout the body. Other **neurobiologists**

✴ study how the eyes communicate with the brain.

✴ investigate animals, such as bats, that use sound waves to navigate.

✴ explore the way birds learn songs.

Up close and personal with a coral that Vincent saw in Belize.

Brain Breath

An adult human brain weighs about 1.3 kilograms (3 pounds). Even though that's only 2 percent of an average person's weight, the brain receives a whopping 20 percent of our blood supply! Why? Give your brain a chance to work on the answer, then discuss it with a partner.

Is It 4 U?

What parts of Vincent's job would you like?

- Traveling
- Discovering tools that are used to advance science
- Teaching others about your discoveries
- Advocating for the environment
- Starting a business

Coral Reef Grief

Vincent's dives in coral reefs have made him realize that many of them are endangered. Scientists are only just discovering how reefs benefit humans. What might happen if reefs disappear? Share your ideas with your class.

Check out your answers on page 36.

Overcoming the Obstacle

Miriam credits her success to working hard to overcome her learning disorder and to finding a field that she is passionate about. When she was growing up, Miriam loved fitness. She competed in equestrian events and loved to ski. She liked nutrition because it blended her interest in science with her enjoyment of exercise and fitness.

MIRIAM NELSON

Friedman School of Nutrition
Tufts University

Miriam's Work

As a leading nutritionist, Miriam Nelson has written best-selling health books. She's also appeared on *The Oprah Winfrey Show*, *The Today Show*, and CNN. She has helped thousands of people become healthier through diet and exercise. But becoming successful wasn't easy. Miriam had dyslexia as a girl, and she struggled to read and write. In fact, she was drawn to science and math partly because those subjects had fewer words and more numbers than others.

Getting Stronger, Not Older

Girls and boys develop strong bones and muscles during childhood by playing, running, and jumping—and by eating healthful foods. But as we grow older, our muscles and bones can change and become weaker. Miriam researches the ways simple exercises can reverse those changes. She developed an exercise routine called *StrongWomen* that has shown many women (and men) exercises they can do to become stronger. She also shattered a myth of aging—by showing that even people in their 80s can get stronger by exercising.

Miriam helps a research volunteer stay fit.

A nutritionist is a science professional who studies food and diets. Miriam researches ways that diet and exercise can improve the health of women and the elderly. Other **nutritionists**

✳ work in hospitals to plan diets for patients.

✳ plan meals for schools.

✳ teach healthful eating habits and recommend dietary changes.

Nutrition Nuts and Bolts

Look at the label on a jar, can, or container of store-bought food. Check out the Nutrition Facts section to find the following information.

- How many servings does it contain?
- How many Calories are there in each serving?
- What percentage of those Calories comes from fat?
- How many servings would you eat to consume 2,000 Calories?
- Sodium can cause high blood pressure. If you had two servings, what percentage of the daily value of sodium would you consume?
- Is the food a good source of vitamins? Which ones?

About You

Miriam used to pester her teachers to teach her about things that she was curious about. What kinds of questions do you ask your teachers?

Are U Fruit-full?

Nutritionists recommend we eat at least two servings of fruit each day. How much fruit do you eat?

For one week, keep a journal of the fruit you eat each day. Here are some examples of one serving of fruit.

- 1 medium apple
- 1 medium orange
- 1 medium banana
- ¾ cup 100% fruit juice
- ½ cup fresh fruit chunks
- ¼ cup raisins
- ½ cup grapes or berries

Is your diet fruity enough? Share your journal with a classmate. Discuss why it's important to eat fruit every day.

Edith delivers babies and teaches young doctors how to care for pregnant patients.

Good Challenges

"Medicine is constantly asking you to do more and to learn more. If that's in your personality, medicine is a place for you. It will always ask you to take the next step."

EDITH CHENG

Department of Obstetrics and Gynecology
University of Washington

When I Grow Up . . .

As a girl, Edith Cheng sometimes saw needy people on the busy streets of Hong Kong where she lived, and it made her sad. She decided that when she grew up, she would do something to help people. Edith also decided early on that she wanted to work in science, even though there weren't a lot of female role models for her to follow. She was inspired after doing a book report on the famous physicist Marie Curie. Edith loved that Marie was a pioneer for women in science.

The Stork Revealed

With a lot of hard work and dedication, Edith achieved both of her goals. Today she's an obstetrician—a doctor who takes care of pregnant women and delivers their babies. Edith is no ordinary obstetrician, though. She takes care of women with complicated medical histories or whose children are expected to be born with genetic defects. Her work can be challenging. But Edith says it's rewarding to help a patient through a tough pregnancy and then to "deliver a crying human being with curly hair. It's quite amazing!"

An obstetrician is a doctor who takes care of women's health needs. Edith specializes in caring for patients with complicated pregnancies. Other **obstetricians**

* provide routine medical care for women.
* take care of patients with normal pregnancies.
* conduct research on birth defects and abnormalities.

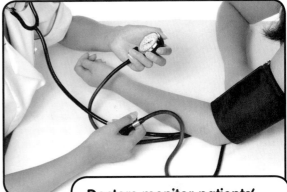

Doctors monitor patients' vital signs—respiratory rate, body temperature, pulse, and blood pressure (above).

Think About

Edith worked as a genetic counselor for several years before she became a doctor. "I think I was afraid that I couldn't do it," she explains. Think about something that you've done that at first you weren't sure you could do. What did you learn about yourself?

Write It Out

In the fifth grade, Edith read a biography of Marie Curie. It made her start thinking about becoming a scientist. Pick a partner and take turns discussing the people who inspire you.

Vital Stats

1	Number of television shows on which Edith has appeared—she was featured in the Lifetime TV real-life show *Women Docs*
2	Number of teachers Edith thanks for sparking her interest in science
20	Number of patients Edith sees in one day
1,000+	Number of babies Edith has delivered or helped deliver over her career

Mechanical Interest

Ralph became interested in orthotics after he shattered his leg as a young Army medic. He studied chemistry, physiology, and physics in a two-year training program in orthotics. His education continues today, as computers and technology advance the field. One thing that remains the same is that being an orthotist is still about helping others. "It's a good career," Ralph says.

RALPH URGOLITES

Walter Reed Army Medical Center

A Supporting Role

When soldiers are injured and lose the ability to move parts of their body, Ralph Urgolites is there to help. As an orthotist at one of the country's top military hospitals, Ralph designs and builds supportive neck, arm, or leg braces. These allow patients to move more easily or to regain strength in their muscles.

Ralph runs the hospital's orthotics and prosthetics lab. He also works closely with specialists who make artificial limbs and with the patients who need them. His busy department has more than 1,000 amputee visits each year and has helped countless people to walk again. The work is rewarding and sometimes emotional, Ralph says. "If you want to help people to walk for the first time, really this is the field to be in," he says. "You cry . . . their parents cry. . . ."

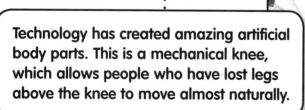

Technology has created amazing artificial body parts. This is a mechanical knee, which allows people who have lost legs above the knee to move almost naturally.

An orthotist builds braces and other supportive equipment for the body. Ralph helps injured soldiers to move more easily. **Orthotists** also

* design braces for people with neck or wrist injuries, including carpal tunnel syndrome.
* teach people to walk or run using prosthetics.
* build footwear for people with foot injuries or deformities.
* outfit children with cerebral palsy or scoliosis with spinal braces.

Advances in the past ten years have led to many exciting new developments, including this "running leg," which lets people exercise and even run marathons.

Boy Mechanic

Growing up, Ralph liked to fix and build things. He took after his dad, who was an airplane mechanic. He also took after his mother, who helped others as a nurse. On a piece of paper

* list some of the things you like to do.
* list some of the people you like to help.

Now, make a third list that combines these two. Include the things you like to do that help other people, too.

Is It 4 U?

Ralph's job includes

* working with veterans injured in war.
* working with children who have illnesses or birth defects.
* using a computer to design braces or limbs.
* working on a team to build limbs and braces.

Which parts of his job would you like doing?

BENJAMIN CARSON

Johns Hopkins Medical Institutions

Learning to Learn

Growing up poor in a single-parent family in Detroit, Benjamin Carson spent most of his childhood angry. Failing at school, Ben felt like he was "the stupidest kid in the fifth grade." Then, Ben's mother laid down new rules. Ben and his brother couldn't watch TV, they had to do their homework, and they had to read and write book reports on two books a week. Soon Ben was reading stories about animals, plants, and rocks, and answering questions at school. He was thrilled. "Once I recognized that I had the ability to pretty much map out my own future based on the choices I made and the degree of energy I put into it, life was wonderful."

Gifted Hands

By the seventh grade, Ben had moved from the bottom of his class to the top. He later earned a scholarship to Yale University. Then he attended medical school. It was there that he discovered his gift for surgery—steady hands and terrific hand-eye coordination. Today Ben is one of the top brain surgeons for children in the world. He performs about 500 surgeries a year. Many of them dramatically improve the lives of his patients forever.

Ben established Carson Scholars, a scholarship fund. Through his organization, he supports academically successful students who give back to their communities.

Ben reviews digital pictures of a brain.

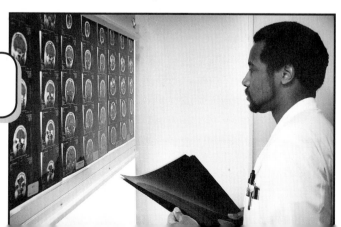

A neurosurgeon is a medical doctor who does surgery on the brain, spinal cord, and nerves. Ben specializes in treating children with brain diseases or deformities of the face and skull. Other **neurosurgeons**

✳ treat patients suffering from back and neck pain.

✳ take care of those at risk for strokes.

✳ treat head and spinal cord injuries.

✳ figure out ways to prevent and treat concussions.

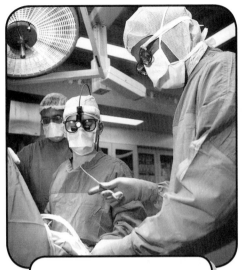

Brain operations often last for many hours.

Think About

Ben made history by performing the first successful surgery to separate twin babies who were joined at their heads. Taking risks is a big part of Ben's work. "There is almost nothing in science that we do perfectly from the beginning," he says.

Divide a sheet of paper into three columns with these labels.

1. *What I Tried*
2. *How I Struggled*
3. *What I Learned*

Describe something you've tried to do, whether you succeeded or not. Tell how you struggled, and what you learned from it.

Surgical Tools

Being a successful brain surgeon requires many skills. Read the list below and discuss with a partner which characteristics you share with Ben.

- Knowledge of biology
- Curiosity
- Ability to concentrate for long periods of time
- Hand-eye coordination
- Ability to make decisions
- Ability to remain calm under pressure

JULIE O'CONNELL
AthletiCo at East Bank Club

Gold Medal Effort

Julie O'Connell lived up to her reputation as a physical therapist for the U.S. Women's Soccer Team during the 2004 Summer Olympics. Two days before the final match, soccer star Julie Foudy sustained a foot injury. It might have taken four to six weeks to heal, but Julie O'Connell quickly diagnosed the injury. Then, both Julies "worked tirelessly for two days to get her to recover well enough to play." The effort and Julie's recommendations of exercise, massage, electrical stimulation, and realignments worked. The soccer player was on her feet for the final game—the team won Olympic gold. "It makes all of our hard work worthwhile when you see results like that," Julie says.

The Best for the Best

Julie's patients call her the "Miracle Worker" because she's so good at fixing them when they're injured. That's high praise, considering that Julie works with some of the best athletes in the country.

"Julie has the most amazing feel for the body and how joints should operate. I've become completely dependent on her!"—Julie Foudy

A physical therapist

is a health professional who uses exercise and other treatments to heal injured muscles, tendons, and ligaments. Julie works with top athletes. Other **physical therapists**

✳ work with people learning to walk again.

✳ help people get into shape through exercise.

✳ relieve pain in people who've been in accidents.

✳ plan workout programs for sports teams.

Is It 4 U?

Check out this list of some of the things physical therapists do.

- Teach people how to get in shape
- Diagnose sports injuries
- Evaluate how certain injuries occur
- Treat athletes' injuries during competition
- Research ways to prevent sports-related knee, ankle, and hip injuries

What parts of being a physical therapist would you enjoy doing? Team up and discuss them with a partner.

Funny Bones

Q. How many knees do you have?

A. Four—your right knee, your left knee, and your two kid-neys.

Helpful Experience

Julie says a good physical therapist knows how an athlete is using his or her muscles. Julie understands soccer players because she played the game in high school. What sports do you play? What muscles do you use?

Deborah now teaches at the Harvard School of Public Health.

DEBORAH PROTHROW-STITH
Harvard School of Public Health

Asking Questions

As a medical student working in the emergency room, Deborah Prothrow-Stith once stitched up a young man who had been in a knife fight. It made her think. Doctors teach people how to avoid dangers like heart disease and lung cancer. Why don't they teach people how to avoid violence?

Going Beyond the E.R.

Deborah wanted to do more than "stitch 'em up, send 'em out." So she began to research how living near violence or being the victim of violence affects people's lives and their health. Deborah went on to create some of the first programs in the nation to teach teenagers how to avoid violence and change their attitudes toward it. Today many schools have adopted programs based on her work.

Starting Strong

Deborah also helped to start a type of healthcare called public health practice. It focuses on encouraging people to adopt healthful habits and to prevent illness in the first place.

Deborah's book suggests ways to stop girls' violence.

A professor of public health studies and teaches about health issues that affect the general public. Deborah's research focuses on preventing violence. Other **public health professionals**

* stop diseases from spreading.
* test for environmental pollution.
* educate people about how to prevent illnesses.

"You don't have to be brilliant, you just have to work hard and be committed."

You R What U Eat

It's true! Scientific studies show over and over again that our health is affected by what we eat. Keep a food journal for a week. Sort the foods you eat into the five basic food groups—grains, vegetables, fruits, milk, and meat and beans. Then, based on the chart, plan a healthful menu of school lunches for the next week. Bon appetite!

6 oz.
Grains

1½ cups
Fruit

3 cups
Milk

5 oz.
Meat and Beans

2½ cups
Vegetables

Teenagers who consume 1,800 Calories per day need these amounts from each food group. You may need to adjust the amounts so they're right for you.

Q and A

Public health professionals interview people to learn more about their health issues. What questions would you ask students at your school to find out how they handle conflict and anger?

* Make a list of questions you would ask.
* Compare your list with a classmate's list.
* Put your cool heads together and brainstorm positive ways to deal with conflict and anger.

Healthful Habits

The goal of Deborah's type of healthcare is to prevent illness and injury by encouraging healthful habits. What are your healthful habits? Make a list of them, and then make a list of your *un*healthful habits. Now, what can you do to change each *un*healthful habit so you can cross it off the list?

About Me

The more you know about yourself, the better you'll be able to plan your future. Start an **About Me Journal** so you can investigate your interests, and scout out your skills and strengths.

Record the date in your journal. Then copy each of the 15 statements below, and write down your responses. Revisit your journal a few times a year to find out how you've changed and grown.

1. *These are things I'd like to do someday.*
   Choose from this list, or create your own.

   - Invent treatments to improve people's lives
   - Conduct experiments in space
   - Study chemical reactions
   - Design and build artificial limbs
   - Report medical news
   - Treat children who are ill or injured
   - Help stop an outbreak of disease
   - Fix athletes' injuries
   - Work at a large hospital
   - Illustrate medical posters
   - Treat complicated illnesses

2. *These would be part of the perfect job.*
   Choose from this list, or create your own.

   - Being outdoors
   - Making things
   - Writing
   - Designing a project
   - Observing
   - Being indoors
   - Drawing
   - Investigating
   - Leading others
   - Communicating

3. *These are things that interest me.*
   Here are some of the interests that people in this book had when they were young. They might inspire some ideas for your journal.

   - Building and fixing things
   - Doing science experiments
   - Writing book reports
   - Studying famous scientists
   - Drawing
   - Reading about different scientific discoveries
   - Working with numbers
   - Riding horses
   - Skiing
   - Exercising
   - Asking questions
   - Helping people
   - Working on computers
   - Playing soccer

4. *These are my favorite subjects in school.*

5. *These are my favorite places to go on field trips.*

6. *These are things I like to investigate in my free time.*

7. *When I work on teams, I like to do this kind of work.*

8. *When I work alone, I like to do this kind of work.*

9. *These are my strengths—in and out of school.*

10. *These things are important to me—in and out of school.*

11. *These are three activities I like to do.*

12. *These are three activities I don't like to do.*

13. *These are three people I admire.*

14. *If I could invite a special guest to school for the day, this is who I'd choose, and why.*

15. *This is my dream career.*

Careers 4 U!

Medical Sciences
Which ^career is 4 U?

What do you need to do to get there? Do some research and ask some questions. Then, take your ideas about your future—plus inspiration from scientists you've read about—and have a blast mapping out your goals.

On paper or poster board, map your plan. Draw three columns labeled **Middle School**, **High School**, and **College**. Then draw three rows labeled **Classes**, **Electives**, and **Other Activities**. Now, fill in your future.

Don't hold back—reach for the stars!

Exercise Physiologist

Veterinarian

Orthopedic Surgeon

Medical Illustrator

Space Nutritionist

Behavioral Biologist

Orthodontist

Biology Teacher

Biomedical Engineer

Pharmacist

Nurse

Geneticist

Environmental Health Scientist

Psychologist

Surgeon

Forensic Science Technician

Respiratory Therapist

Immunologist

Optometrist

Microbiologist

Epidemiologist

Physical Therapist

Anesthesiologist

Neurologist

Pediatrician

Chemist

Glossary

amputee (n.) A person who has had an arm, leg, or finger removed, usually because it is diseased or damaged. (p. 22)

biology (n.) The study of living things. It includes the study of how plants, animals, and microorganisms develop, live, reproduce, and interact with their environment. (pp. 12, 14, 25)

calorie (n.) The amount of heat energy needed to raise the temperature of 1 gram of water 1 degree Celsius—also the amount of heat energy that 1 gram of water releases when it cools down by 1 degree Celsius. A Calorie is actually a kilocalorie and is used to indicate the energy content in food. (A kilocalorie is 1,000 calories.) (p. 19)

chemistry (n.) The study of the elements and the ways in which they interact with each other. (p. 9)

diabetes (n.) A disease in which there is too much sugar in the blood. A person with diabetes either cannot make enough insulin, the chemical compound that cells need to take in sugar properly, or cannot use it effectively. (p. 8)

dietary (adj.) Having to do with the typical food and beverages ingested by a person. (p. 19)

epidemic (n.) When many people suffer from the same disease at the same time, it is called an epidemic. Some diseases only occur in certain parts of the world (endemic), other diseases occur all over the world (pandemic). (pp.11, 14)

genetics (n.) The study of how living organisms, including people, inherit traits or genes from their parents or parent. (pp. 20, 21)

immune system (n.) The system that protects the body from infection by microorganisms and disease, and includes the skin and the respiratory, digestive, and circulatory systems. (p. 6)

infection (n.) A disease that is caused by microorganisms entering the body. (p. 9)

infectious (adj.) Describes a disease that can be spread from person to person when microorganisms enter the body through the air or by contact. (p. 11)

journalist (n.) A person who writes for a newspaper or magazine and presents the facts or events without interpretation. (p. 14)

nervous system (n.) The organ system made up of the brain, spinal cord, and nerves that controls and communicates with other body systems. (p. 16, 17)

physician (n.) A person who is trained and licensed to treat illness and injuries, and to teach people how to prevent them; also called a doctor. (p. 7)

prescription (n.) An order written by a medical doctor to a pharmacist that prescribes medicine for a patient. (p. 7)

prosthetics (n.) The branch of medicine or dentistry that deals with the replacement of missing body parts with artificial ones. (pp. 22, 23)

respiratory system (n.) The system that provides your body's cells with oxygen and removes carbon dioxide waste. (pp. 10, 21)

surgeon (n.) A medical doctor who specializes in the removal or repair of diseased or damaged parts of the body. (pp. 6, 24, 25)

surgery (n.) 1. The branch of medicine that deals with the removal and repair of diseased or damaged parts of the body. 2. An operation performed by a surgeon. (pp. 24, 25)

virus (n.) Nonliving particles smaller than bacteria that are able to multiply only inside living cells. Viruses cause many diseases, such as the flu, measles, and the common cold. (p. 11)

weightlessness (n.) The condition in which all objects float in space. When an object (or astronaut) is weightless, it seems as though it is not subject to Earth's gravitational pull. But, in fact, it is gravity constantly pulling the object toward the center of Earth that keeps it in orbit. (pp. 6, 7)

Index

CHECK OUT YOUR ANSWERS

ASTRONAUT, page 7

Around and Around

- On an eight-day mission you would orbit Earth 128 times.
- Each day, you would travel 672,000 kilometers (about 420,000 miles).

EPIEMIOLOGIST, page 11

You, the Epidemiologist

Here's one hypothesis. Since the grill wasn't working well, it might not have reached a high enough temperature to kill bacteria in the chicken. Even healthy chickens might carry bacteria that can make people sick. To kill bacteria, all chicken must be cooked until the temperature inside the chicken is at least 82°C (180°F).

NEUROBIOLOGIST, page 17

Brain Breath

Blood carries oxygen from the lungs and nutrients from digested food to the brain. The brain's 100 billion nerve cells are constantly active so they need a constant supply of energy (nutrients) and oxygen.

Coral Reef Grief

If coral reefs disappear
- important fish habitats will be lost.
- coastal cities and towns will be more vulnerable to destruction by storms.
- many species of ocean life in the coral reef ecosystem may lose their homes.
- tourism will decline in areas where coral used to thrive.

IMAGE CREDITS

Dmitriy Shironosov: Cover. Corbis: pp. 2-5 and pp. 30-31 background. NASA: p. 2 (Seddon), p. 5, p. 6, p. 7. Karen Sideman: pp. 6-29 top banner. Courtesy Ronald E. Eastabrook: p. 2 (Serabjit-Singh), p. 8. Courtesy Simon and Shuster: p. 2 (Kolata), p. 14. Courtesy Richard Chase: p. 3 (Prothrow-Stith). Sally Ride Science: p. 4. Joanne Kim: p. 9. Centers for Disease Control: p. 11, p. 32 right, p. 33 bottom right. Donald Bliss: p. 12, p. 13. Andrew Brucker: p. 14 bottom. Bernhard Lelle: p. 17. Marc Dietrich: p. 19. Courtesy White House: p. 23. Shawn Pecor: p. 27. Courtesy of Richard Chase: p. 28, p. 29. Courtesy John B. Wiley & Sons: p. 28 bottom. Clara Lam: p. 30. Fernando Audibert: p. 33 top left. Jyn Meyer: p. 33 bottom left.

Sally Ride Science is committed to minimizing its environmental impact by using ecologically sound practices. Let's all do our part to create a healthier planet.

These pages are printed on paper made with 100% recycled fiber, 50% post-consumer waste, bleached without chlorine, and manufactured using 100% renewable energy.